Roberta Mertz Rodrigues
Sonia Tomie Tanimoto
Ana Paula Machado Velho

Phytoremediation using ornamental plants

Roberta Mertz Rodrigues
Sonia Tomie Tanimoto
Ana Paula Machado Velho

Phytoremediation using ornamental plants

for the recovery of urban areas contaminated with lead

ScienciaScripts

Imprint

Cover image: www.ingimage.com

This book is a translation from the original published under ISBN 978-620-2-17532-6.

Publisher:
Sciencia Scripts
is a trademark of
Dodo Books Indian Ocean Ltd. and OmniScriptum S.R.L publishing group

120 High Road, East Finchley, London, N2 9ED, United Kingdom
Str. Armeneasca 28/1, office 1, Chisinau MD-2012, Republic of Moldova, Europe
Printed at: see last page
ISBN: 978-620-7-97098-8

Table of contents:

DEDICATORY

I dedicate this work to all those who contributed to its completion.

ACKNOWLEDGMENTS

To God for the perfection of this world and the fluidity of life.

To the University Center of Maringá, for making it possible for me to do this work.

To the Coordination for the Improvement of Higher Education Personnel (CAPES), for the scholarship.

To the professors of the Postgraduate Program in Clean Technologies/ Unicesumar for their willingness to share their precious knowledge.

To my advisor Prof.ª Sonia Tomie Tanimoto, for her innovative research ideas, her ability and ease to permeate all areas and all the support she provided.

To the laboratory technician Marcelo Teixeira for his support, patience and guidance in laboratory methods, without which the analysis would not have been possible.

To Prof. Djalma Palin Junior and his company Termo Scientific for their willingness to carry out the chemical analysis.

To all my friends who, in one way or another, supported me throughout this work.

Finally, I would like to thank my parents, Wilson and Clarice, from the bottom of my heart, for all their support during this stage of my life, without which none of this would have been possible.

"Life doesn't know time, it knows flow".

Ernst Götsch

Phytoremediation using ornamental plants to recover urban areas contaminated with lead

SUMMARY

The main objective of this study was to evaluate the resistance of ornamental plants in environments contaminated with Pb (lead) and to determine whether they can be used as lead remediators in the environment. To this end, an investigation was carried out into the remediation potential of the species *Tradescantia pallida* (purple ragwort), *Ophiopongo jaburani (*snake beard), *Sansevieria trifasciata (*St. George's sword) and *Cuphea gracilis* (erica or false erica) in absorbing/adsorbing lead from artificially contaminated soils. The seedlings were transplanted into plastic pots containing substrate and soil. The experiment was laid out in a completely randomized design and a quantity of approximately 20.0 grams of lead sulphate II (PbSO4) was added to the soil. During the experiment, weekly photographs were taken to monitor and compare the development of the plants. After 20 weeks of contamination, the plants were prepared for chemical analysis using flame and plasma atomic absorption spectroscopy. The results show that all the species studied have the potential to adsorb Pb on their roots. When visually evaluating the growth of the species, comparing the uncontaminated and contaminated individuals, it was not possible to see any signs of intoxication, as they had "normal" growth. With the exception of *S. trifasciata*, plants develop defense mechanisms to prevent the contaminant from being translocated to the aerial part. All the species studied that received Pb contamination showed a similar appearance to the control plants and none of the individuals died, indicating resistance to the element.

Keywords: Pb; phytoextractor; heavy metals; phytoindicator; phytostabilization; remediation

Chapter 1
1 INTRODUCTION

Cities are living and complex organisms, as well as being open ecosystems, built on top of the natural environment which artificialize the environment to suit human needs. The greater the order of artificialization, i.e. the technological elaboration of the goods produced/used and discarded, the greater the environmental impact.

According to Bruna (2002), industrialization took place in Brazil from the 1960s onwards, also intensifying the phenomenon of urbanization. Added to this, the increase in life expectancy and the decrease in mortality has increased the population. Brazil's current population is 165,371,493, with an urbanization rate of 78.4% (IBGE, 2015).

Cities find it difficult to absorb all these individuals in a planned way. This leads to irregular occupation of valley bottoms, hillsides and protected environmental areas. The lack of basic sanitation and domestic sewage treatment aggravates this situation, along with the need to get around, which will lead to environmental degradation, polluting the soil, air and surface and groundwater (BRUNA, 2002).

With the aim of regulating and minimizing the impacts caused by accelerated urbanization, the National Solid Waste Policy (PNRS) came into force in 2012. It prohibits the establishment of new garbage dumps and includes resolutions on the implementation of sanitary and controlled landfills capable of correctly disposing of different types of waste (BRASIL, 2010).

However, the Brazilian Association of Public Cleaning and Special Waste Companies (ABRELPE) (2013) points out that the panorama of solid waste in Brazil quantifies the generation of urban solid waste (MSW) at 209,280t/day, resulting in an average of 1.041kg/inhab/day, with 189,219t/day or 0.941kg/inhab/day being collected, reaching a figure of 20,000 tons per day being disposed of inappropriately. However, only 58.26% of MSW in Brazil is properly disposed of and just over 62% of municipalities have registered some sort of selective collection initiative.

In addition to the problem of solid waste and its disposal, there is widespread concern in cities about the contamination of their spaces, since some toxins that pose a high risk to the health and survival of an entire ecosystem can be released irresponsibly and without any treatment by various types of industries and services.

Within this parameter, we can highlight soil contamination by lead (Pb), a risk generated by industries and services that use this element in their production process. According to Sharma and Dubey (2005), the sources of environmental contamination by Pb are automobile exhaust, pigment additives and gasoline; mining, lead ore smelting industries through factory chimneys; effluents from the accumulator battery industry; metallic paints and finishing operations; fertilizers and pesticides; as well as natural weathering.

In the region, we can highlight the city of Umuarama, which has an automotive battery recovery plant and a branch in Cianorte. In Maringá, there are industrial laundries, which according to the Clothing Industries Union (Sindvest) are divided into six industrial laundries and twelve print shops, as well as three paint and varnish factories operating in the city.

Contamination by this material is rare, but when it does occur, it has a major environmental and economic impact and, above all, affects people's health. These contaminated areas are considered a major risk to public health and safety, which limits local urban development and devalues the entire real estate sector (SÁNCHEZ *apud* GLOEDEN; SOARES, 2013).

In this way, this research was developed in the search for functional alternatives for the potential recovery of Pb-contaminated soils, either globally or partially, making them suitable for use without health risks. Prioritizing characteristics such as low cost and maintenance, maintaining efficiency and considering the speed of the remediation process.

Phytoremediation has been adopted, which is a method that uses plants to degrade, extract, contain or immobilize contaminants in soil or water. Much of this method is based on empirical knowledge in agriculture, forestry and horticulture regarding environmental issues; contaminants can be of various types: organic and inorganic (EPA, 2000).

The main advantages of this method are its wide variety of plants, its reach, its efficiency and the possibility of using it in different environments (soil, water) and, above all, its ability not to harm the environment (MALIK; BISWAS, 2012).

Considered a green and sustainable technology, it does not require the implementation of energy or intensive monitoring; it improves air quality by sequestering carbon dioxide, controlling erosion, runoff, infiltration and atmospheric dust. It is characterized by being a passive, on-site treatment of contamination; it favors public perception and educational opportunity. It improves aesthetics and helps reduce noise; it is applicable to remote sites that are difficult to access; it can be used in conjunction with other remediation methods. It is suitable for identifying and mapping contamination. It can be applied as a preventative to detect possible leaks; with low maintenance that helps nature to recover; creates habitats; promotes natural restoration and is financially viable (EPA, 2000).

Therefore, based on the fact that some plant species are able to absorb/adsorb contaminants in the soil for an indefinite period, the main objective of this study is to evaluate the resistance of ornamental plants in lead-contaminated environments and determine whether they can be used as remediators of this metal in the soil. In addition to analyzing the potential of the species *Tradescantia pallida* (purple ragweed), *Ophiopongo jaburani (*snake beard), *Sansevieria trifasciata (*St. George's sword) and *Cuphea gracilis* (erica or false erica) to absorb/adsorb lead from contaminated soils; to visually monitor the growth of the species by comparing uncontaminated and contaminated individuals, observing possible signs of poisoning; to determine the remediation potential of each plant through chemical analysis and, finally, to determine the presence of lead in the different regions of the plants (roots, leaves, stems).

Chapter 2
2 LITERATURE REVIEW

3 .1COMMERCIAL IMPORT OF LEAD

Lead, with an atomic number of 82, is a bluish-gray metal, malleable and sensitive to air. Its melting point is relatively low at 327°C and its boiling temperature is 1749°C. It is classified as a heavy metal with high toxicity. It is relatively abundant in the earth's crust, with the largest natural sources being volcanic emissions, geochemical weathering and water mists (WHO, 1995). Known since antiquity, it is the fifth most used metal in industry and can be extracted from mines and transformed by various processes. In the last 100 years, production and consumption have practically quadrupled (PANTAROTO; FIGUEREDO, 2007).

According to the National Department of Mineral Production (DNPM) (2014), in 2012, world lead production reached 10.5 million tons, with China having the largest share of 44%. There is no production of primary lead in Brazil, the handling of this element is restricted to its recycling; it is found mainly in: automotive batteries, industries and telecommunications materials. Even so, in 2013, the country managed to recycle a total of 152,000 tons, with refining plants having a capacity of 160,000 tons/year (DNPM, 2014).

Commercially, it is widely used due to its vast application in industry and its peculiar characteristics, such as its exceptional malleability, low melting point, high resistance to corrosion, high density, high opacity to X-rays and gamma rays, electrochemical reaction with sulphuric acid and chemical stability in air, soil and water. It can also be used in pure form or alloyed with other metals or chemical compounds (ATSDR, 2007).

2.2 1,'TII.I/ACTION OF LEAD AND CONTAMINATED URBAN SOIL

In the 19th century, metallic lead was used to seal food and drink cans, but when opened, the lead in the solder reacted with oxygen and there was a risk of contaminating the food, especially in acidic environments, which are widely used as preservatives (SPIRO; STIGLIANI, 2009, BAIRD; CANN, 2011).

Baird and Cann (2011, p. 80) report that "[...] the Franklin Expedition of 1845, where the aim was to find a passage in the Arctic, failed because all its members died of lead poisoning from the solder in the tin cans that carried the food".

It is important to note that dissolved lead (Pb^{2+}) is completely absorbed by the organism;

today it is still widely used to make ammunition used by poachers and can affect local fauna and fishing items, directly contaminating aquatic organisms (SPIRO; STIGLIANI, 2009, BAIRD; CANN, 2011).

According to Yanakiew (2005 *apud* OLIVEIRA; GOMES; AFONSO, 2010, p. 242) "[...] around 70% of the heavy metals in US landfills come from electronic waste, with around 40% by mass of the lead found in the country's landfills coming from electronic equipment".

In paints and pigments, lead is mixed in the form of salts because they are excellent fixatives. These include chromates ($PbCrO_4$) which form yellow pigments and are widely used in horizontal traffic signs, oxides (Pb_3O_4) with a red color, the basis of anti-corrosion paints for metal structures and hydroxycarbonates [$Pb_3(OH)_2(CO_3)_2$] known as white lead, widely used as the basis for internal paints with white pigmentation (WHO, 1995, SPIRO; STIGLIANI, 2009, BAIRD; CANN, 2011).

The levels of lead in paints have already been reported on television programs, such as in a report by Fantástico in 2015, where alarming results were presented following an analysis by the National Institute of Metrology, Quality and Technology (INMETRO), which rejected two brands of synthetic enamel for having amounts of lead higher than those permitted by national legislation, one of which had an amount 200 times higher than that permitted.

Similarly, a report in the newspaper O Globo on studies carried out by the University of California and the Berkeley School of Public Health found that of the 32 types of lipstick and gloss on the American market, at least half contained lead in quantities above what is permitted in the country. In Brazil, an analysis by the Chemistry Department of the Federal University of Minas Gerais showed that of 22 types of lipstick on the Brazilian market, only one, of French origin, did not contain lead (MILHORANCE, 2013).

Lead is also used as a ceramic additive in the glazing process. It is fused to the surface of the ceramic to make the material water-resistant and give it a polished finish (SPIRO; STIGLIANI, 2009).

It began to be used in gasoline in 1922 to increase the fuel's performance and octane rating. Its addition only began to be reduced in 1990 by some countries that opted for the use of catalysts (SPIRO; STIGLIANI, 2009, BAIRD; CANN, 2011).

As the automobile industry grew in the 20th century, so did the sale of batteries, which are responsible for starting, lighting and igniting cars. It is estimated that 70 to 75% of the world's lead production is used for batteries. Batteries are electrochemical devices that provide electrical energy through chemical reactions, and lead-acid batteries have the advantage of working with reversible chemical reactions, which allows them to be recharged (PAOLIELLO; CHASIN, 2001).

To recycle lead-acid batteries, three processes are carried out: battery breaking, reduction and lead refinement. In the most modern processes, battery breaking has virtually no human contact, using mechanized and confined technologies, but can generate environmental impact through dust, particulates and lead debris (JOST, 2001 *apud* PAOLIELLO; CHASIN, 2001).

Battery scrap usually consists of metallic lead, lead oxide, lead sulphate and other metals such as calcium, copper, silver, antimony, arsenic and tin. The reduction process involves isolating the metallic lead. At this stage, there is a possibility of contamination from the disposal of contaminated waste, both lead dust and chlorinated compounds and sulphur dioxide (SO_2) (JOST, 2001, *apud* PAOLIELLO; CHASIN, 2001).

The purpose of refining is to separate out the remaining copper, antimony, arsenic and tin, transforming hard lead into soft lead. Here the environmental impacts can include the emission of lead vapors, sulfur dioxide (SO_2), heavy metal dust and chlorine gas (JOST, 2001 *apud* PAOLIELLO; CHASIN, 2001). Generally, the remnants of refined lead remain close to the polluting source, however, it is hypothesized that 20% is dispersed to more remote areas, depending on the size of the particulate matter, generated mainly by smelters, incinerators and the burning of fossil fuels (WHO, 1995).

Lead can be transported from the atmosphere to the environmental surface through dry or wet deposition, with wet deposition being more worrying as it accounts for between 40 and 70% of the dispersions that cause lead to accumulate in the soil. Particulate transport also depends on a number of factors including pH, the mineral composition of the soil and the amount and type of organic matter (REAVES; BERROW. 1984. GARCIA-MIRAGAYA, 1984 *apud* WHO, 1995).

The solubilization of organic lead is directly related to high pH, with its most soluble form being between 4 and 6 and available to plants and groundwater (EPA. 1986 *apud* WHO. 1995). The highest concentrations of lead in the soil are always close to the smelting and battery recovery industries, as well as highways with heavy traffic in large urban centers (WHO. 1995).

A current case of contamination occurred in Flint, a US city, when it changed its water supply from Lake Huron to the Flint River. However, the Flint River is very acidic, which reacts with the lead pipes and corrodes them, contaminating people in their own homes (NATIONAL JOURNAL, 2016).

In Brazil there are cases of soil contamination by lead, caused by lead mining. The city of Santo Amaro da Purilicacao in the state of Bahia is considered one of the most lead-polluted cities in the world, according to studies by the Federal University of Bahia. It is estimated that 80% of the population is contaminated by the waste left behind by the mining company Pumbum Comércio e Representacóes de Produtos Minerais e Industriais, formerly Companhia Brasileira de Chumbo - COBRAC, which processed minerals and operated in the town from 1960 to 1993. The tragedy, which began in the 1960s, is still unfolding today, with the damage caused to the environment resulting in the contamination of the population. Without a management plan to close down the mining company, the population, unaware of the risks, went so far as to pave the streets, build houses and even school buildings and nurseries with the lead waste used in the industrial activity (FERNANDES; BERTOLINO; EGLER, 2012).

Another area explored by the same mining company was the Upper Ribeira Valley, located in the extreme northeast of the state of Paraná and the southeast of the state of São Paulo. Mining for lead, zinc and silver operated until 1996, causing a major impact on the vegetation and landscape, and the processing and refining of the metals produced mountains of tailings that are exposed. Near the bank of the Ribeira River is the municipality of Adrianópolis-PR with a population exposed to contamination (LOPES et al., 2006).

A recent catastrophe was the collapse of the Fundao dam, which caused the Santarém dam of the Samarco mining company to overflow in Mariana, in the state of Minas Gerais, on November 5, 2015. The collapse released around 60 million cubic meters of mud, which devastated the municipality of Marina, traveled approximately 700 km along the Rio Doce and flowed into Espírito Santo, generating a major environmental impact (COSTA, 2015). A week after the dams broke, the Autonomous Water and Sewage Service of Baixo Guandu, in Espírito Santo, released analyses of the water in the Rio Doce. It showed high quantities of heavy metals such as arsenic, lead, chromium, zinc, barium and manganese, requiring the suspension of water abstraction to supply the population, in addition to having eliminated the river's biodiversity (CONSTANTI; MENEZES, 2015).

2.3 HUMAN LEAD POISONING

In the human body, lead has no function as a nutrient, however, it is practically impossible to find individuals without some significant amount of this metal in their bodies (TROSTER, 1993).

Lead, when found in inorganic form, can affect systems in the human body, causing disorders, mainly affecting the bones. Its action varies depending on the exposure time and intensity, as well as the characteristics of the individual (MOREIRA; MOREIRA, 2004, HOLZBACH et al., 2012). When Pb is in organic form, tetramethyl and tetraethyl, it is fat-soluble and easily absorbed by the body, causing neurological disorders (TEIXEIRA et al. 2004 *apud* HOLZBACH et al., 2012).

The human body absorbs lead slowly and gradually and its accumulation can lead to poisoning. For this reason, it is extremely important to replace this element with less aggressive compounds, both in the formulation of gasoline, using aromatic compounds, and in the pigmentation of paints using itrium hydroxide (SPIRO; STIGLIANI, 2008).

According to Troster (1993), the main sources of human contamination are through the air: in large urban centers, near high-traffic roads and where there is a concentration of industries, and in rare cases through water: mainly after the replacement of lead pipes in homes, and can also be seen in soft water (low calcium and magnesium content). The greatest occurrence, however, as reported by Who (1995) is through the ingestion of food and drink contaminated by lead present in the air, soil and water, with the highest concentration in contaminated plants being in the roots and the lowest in fruit and seeds, pointing out that industrialized food has an even higher lead content.

Lead poisoning can damage the central and peripheral nervous system and in many cases the motor system, worsening into encephalopathy[1] , both in adults and children. Even in children, this contamination is capable of affecting physical growth and reducing height, causing cognitive impairment and dysfunction of fine sense-motor perception, as well as seriously damaging the optic nerve and auditory system even in adults (MOREIRA; MOREIRA, 2004).

The main problem in the hematological system is anemia, due to the toxic effect on red blood cells in the bone marrow. Lead hinders the body's production of hemoglobin, compromising many enzymatic reactions necessary for the synthesis of heme. Lead thus binds to hemoglobin, preventing the transport of iron within the body (WHO, 1977).

It also damages the renal system and patients often suffer from hypertension. Renal dysfunctions are generally reversible in children and are characterized by cellular physiological changes. In the case of irreversible[2] nephropathy, the characteristic features are vascular corrosion, tubular cell atrophy or hyperplasia and progressive interstitial fibrosis (MOREIRA; MOREIRA, 2004).

In the gastrointestinal system, colic is an early symptom, but in the case of high levels of lead exposure, several other symptoms are combined, such as abdominal pain, constipation, cramps, nausea, vomiting, anorexia and weight loss (ATSDR, 2007).

There is evidence in clinical studies, in the general population and in occupational cases, that lead reaches the cardiovascular system, causing heart damage and increases in blood pressure at high levels of exposure (MOREIRA; MOREIRA, 2004).

Due to its harmful effect on public health, because it is highly contaminating and with a focus on atmospheric detoxification methods, the importance of analyzing natural alternatives that contribute to the decontamination of a site is highlighted. It is believed that the use of phytoremediation (the process of recovering an area of soil using certain plant species capable of absorbing, in addition to the nutrients necessary for their survival, the toxins present in the area) is the most viable option.

---

[1] Encephalopathy is the term used to define any diffuse brain disease with altered structure or function (DAMIANI, 2013).

[2] Characterized by a gradual reduction in kidney function (MOREIRA; MOREIRA, 2004).

2.4 MONITORING AND REMEDIATION

To remediate a contaminated environment, various types of treatment can be used: physical, chemical or biological. The scope can be partial or complete removal of the contaminant, reduction or bioavailability, minimizing toxicity. When selecting the type of remediation, the effectiveness of the method, resources required, decontamination time, financial investment and other factors must be taken into account (MALIK; BISWAS, 2012).

Physical remediation consists of removing the topsoil, moving the contaminated area, washing the soil, solidifying it or excavating it, so that the environment is cleaned up. The process is quick, but involves a high cost and the risk of contaminating other areas when transporting the contaminated material, and is not suitable for large areas of contamination (CETESB, 2001).

Chemical remediation adds chemicals to the contaminant so that it reacts, becoming less toxic to the environment and not being absorbed by plants, so the pollutant remains in the soil in a less harmful way. It requires periodic applications, specialized equipment and manpower, can alter the structure and biological activities of the environment in small proportions, and is not efficient for treating large areas (MALIK; BISWAS, 2012).

Bioremediation is a method that uses the metabolic process of native or introduced microorganisms to break down contaminants.

It is carried out through a natural process of microbial cultures and biological residues that make the chemical changes (BENTO, 2012).

Phytoremediation is a technological system that encompasses various procedures that are still under development and involves the use of plants to clean up or remediate contaminated soils. Considered environmentally appropriate, this technology is a safe and inexpensive way of removing contaminants (GRATÀO, et al., 2005). The process consists of introducing plant species into the contaminated environment, allowing them to assimilate the toxins in their roots and leaves. This process can be a great ally in the decontamination of heavy metals, pesticides, xenobiotics, organic compounds and toxic aromatic pollutants (CUNNINGHAM; OW, 2010).

The use of plants to clean up contaminated environments is not exactly new. Historically, it is known that around 300 years ago certain species were already used to treat waste water (HARTMAN, 1975 *apud* LAZART, 2000). Even in the 19th century, some species with high levels of metal accumulation in their leaves were documented (BAUMANN, 1885 *apud* LAZART, 2000).

However, this technology is not recommended for all types of contamination, for example, in cases where it is deeper, or the concentration of the compound is very high, phytoremediation alone will not be enough to efficiently decontaminate the soil (CUNNINGHAM; OW, 2010). Furthermore, the remediation potential will depend on the interaction between soil, contaminant and plant, which is directly related to climatic conditions (LASAT, 2000).

Phytoremediation occurs in three basic forms, the first two involving absorption processes and the last, adsorption processes. The first is degradation, when the plant is able to use the contaminants in its metabolic process as a nutrient. The second is extraction, when the plant accumulates the contaminant without altering its properties, in its roots, stems and leaves. And the third is concentration or immobilization, where the pollutant is sequestered by concentrating it near the root, preventing it from being dispersed. In the last two cases, the plant mass must be properly disposed of or recycled. The advantage is that the amount of material to be treated is less than removing a large area of contaminated soil (EPA, 2000).

Phytoremediation can be subdivided into six mechanisms that can be differentiated by the way they work, the characteristics of the soil, the contaminated media (clay soil, sandy soil, surface water, groundwater) and the types of contamination associated with the concentrate available in the soil (EPA, 2000).

Firstly, phytostabilization can be defined as the immobilization of a contaminant in the soil through absorption or adsorption by the root, or precipitation within the root zone of the plant. Using plants and roots to prevent migration through wind and water erosion, leaching and dispersion in the soil. It happens in the microbial and chemical zone of the root, or alteration of the soil environment and chemistry of the contaminant. By modifying the solubility and mobility of the metal or by disassociating organic compounds, it acts through sorption, precipitation, complexation and valence reduction of the metal. It is used and most suitable for reducing concentrations of metals such as arsenic (As), cadmium (Cd), chromium (Cr), mercury (Hg), lead (Pb) and zinc (Zn) (VASSILEV et al., 2004).

It has the advantage of revegetating and restoring the ecosystem, as well as not requiring the removal of soil or the harvesting and removal of biomass. As a disadvantage, the contaminant remains in situ; fertilization and soil remediation may be necessary; it is also important to prevent the plant from transporting the metal to the aerial part; the root zone and contaminated soil must be monitored to control the solubilization and leaching of the contaminant; it is considered an intermediate containment process (VASSILEV et al., 2004).

Phytostimulation or rhizodegradation is the degradation of contaminants, usually organic ones, through the microbial activity present in the root zone. Roots release substances such as sugars, amino acids, organic acids, enzymes and others that increase microbial activity in the rhizosphere, thus increasing soil biodegradation. The activity of the root helps in the process of aeration and humidification of the soil, creating favorable conditions for the degradation of the contaminant. This process is indicated for the degradation of TPHs (total petroleum hydrocarbons), PAHs (polycyclic aromatic hydrocarbons), pesticides, chlorinated solvents and sulphactants (MARIANO; OKUMURA, 2012).

The advantages of this process are that the contaminant is destroyed on site; the contaminant is not absorbed by the plant; the contaminant can be mineralized; and the installation and maintenance costs are low. As a disadvantage, the slow development of the root zone; the combination with other plant activities can influence the results analyzed; the plant needs additional fertilization due to competition for nutrients from microorganisms; the proliferation of non-degrading microorganisms can be stimulated; the carbon source can be consumed instead of the contaminant; and it is limited to the depth of the root (MARIANO; OKUMURA, 2012).

In the process of phytodegradation, the absorbed contaminant is broken down through the metabolic action of the plant or degradation occurs externally through the release of elements that cause transformation. The substance's hydrophobicity, solubility and polarity will affect absorption, as will the species and physiochemical characteristics of the contaminant. It is very suitable for treating organic contaminants such as chlorinated solvents, herbicides, insecticides, explosives and phenols; and for inorganic compounds, nitrates (EPA, 2000).

Its benefit is that it takes place in environments free of microorganisms, with a high level of toxic contaminants for the rhizosphere. Its disadvantage is the possibility of toxic elements being produced through the metabolization of the pollutant, which can make it difficult to identify the metabolic process and prove the elimination of the toxic element (EPA, 2000).

Phytovolatilization is the evaporation of the contaminant through plant transpiration, which can release it into the atmosphere, or a modified form of it, through metabolic processes. It is indicated for organics such as hydrochloric solvents, and inorganics such as selenium (Se) and mercury (Hg) (MARIANO; OKUMURA, 2012).

This technique has the advantage of transforming compounds into a less toxic form and, after being released into the atmosphere, can be degraded more quickly than in the soil. However, the toxic elements can be released and the contaminant or its derivative can accumulate in the vegetation, subsequently causing transfer to the fruit (MARIANO; OKUMURA, 2012).

Rhizofiltration occurs by absorption or precipitation of the pollutant into the roots of the plant or by adsorption in the root zone, immobilizing or accumulating it, which can be removed later. It is suitable for surface water and effluent treatment, generally at low concentrations, and is not suitable for soil treatment. Both aquatic and terrestrial plants can be used, with terrestrial plants needing more support; it is a system used both at the site of contamination and elsewhere (CETESB, 2001).

Disadvantages include the constant need to correct the pH of the solution; the influence of chemical reactions in the root still needs to be known; control of the effluent flow is necessary for the treatment to be effective; in addition, the plants need to be grown in greenhouse conditions; the plants need to be disposed of frequently; and there are still differences in the results of laboratory and field analyses (EPA, 2000).

Finally, there is phytoextraction, which occurs through the absorption of the contaminant by the plant's roots, which is then stored or transported and accumulated in the aerial part. In this process it is necessary to harvest the plant that retains the contaminant, but a much smaller mass is left to be disposed of than by digging up the soil. It is a technology that is often applied to heavy metal contamination and is highly recommended for use in soils, sediments and sludge, and can also be used in small areas of water contamination (EPA, 2000, GRATÁO et al., 2005).

The advantage is that the contaminating material can be recovered if it is not toxic. The disadvantages are that hyperaccumulators are slow-growing, have little biomass and a small root system; the biomass must always be harvested and removed and disposed of correctly; the metals absorbed can have a toxic effect on the plants and, furthermore, the results presented in the field have been shown to be inferior to those presented in the laboratory (EPA, 2000, GRATÁO et al., 2005).

Thus, the advantages of phytoremediation can be summarized as the reduced cost compared to other traditional soil remediation methods. Treatment always takes place at the site of the contamination, avoiding possible loss to other areas through transportation. In most cases, it does not generate large volumes of waste, and the biomass can be used for energy generation, animal feed (if it is not contaminated by a toxic element), and as a source of mineral replenishment for areas with a nutritional deficit. A wide variety of plants can be used. It helps to correct the ecological balance of the affected area. This technique is widely accepted by the population because it is natural. It drastically reduces the volume of contaminants for final disposal (EPA, 2000, LAMEGO; VIDAL, 2007, CATROGA, 2009, LAZART, 2000).

Some aspects condition the application of phytoremediation, taking into account that the process depends on the development of living organisms and can be influenced by climatic variations such as rainfall, sudden changes in temperature, solar intensity and strong winds. For this reason, the choice of species is of fundamental importance, paying attention to the speed of growth, the biomass generated and the metabolic processes. The ability to act is restricted to the root range of the chosen species. The plants need to be properly disposed of so that they do not run the risk of being contaminated by the plant mass (CATROGA, 2009, LAZART, 2000).

Periodic monitoring of affected areas is of fundamental importance, so that the risk of toxins being released into the atmosphere can be contained. It is also necessary for the contaminant to be available in the soil for the plant to absorb; if the contaminant is translocated to the aerial part, it could contaminate the food chain. It should be noted that studies in this area are still too limited to draw firm conclusions about its application (EPA, 2000, LAMEGO; VIDAL, 2007). Most species that survive in toxic environments have the ability to avoid heavy metals or are hyperaccumulators in their tissues (KHAN et al., 2000).

Hyperaccumulator plants, by definition, have the ability to accumulate 100 times more metal than ordinary species. While normal accumulator plants have a metabolic need for micronutrients of less than 10ppm in dry matter, hyperaccumulator plants, for example, can accumulate quantities of 10ppm for Hg, 100ppm for Cd, 1,000ppm for Co, Cr, Cu and Pb and 10,000ppm for Ni and Zn in dry mass. These plants do not prevent the entry of metals through the root and have developed specific mechanisms to detoxify the high levels of metal accumulated in the cells, thus allowing the bioaccumulation of high concentrations of metals (LAZART, 2000).

However, most hyperaccumulators have slow growth and low biomass, characteristics that limit the use of these species for cleaning up contaminated soils (CUNNINGHAM; OW, 2010, LASAT, 2000). When using the phytoextraction technique with hyperaccumulators, it is necessary to subsequently harvest the plant, which stores the contaminant in its tissues (LAMEGO; VIDAL, 2007).

It was at the end of the 19th century that the high presence of Zn was observed in the species *Thlaspi caerulescens* and *Viola calaminaria,* when grown in soils rich in the metal. This discovery led to research into hyperaccumulator plants. Currently, more than 400 metal-accumulating species have been identified, belonging to 45 families. The vast majority of them are capable of accumulating only one type of metal, but there are multi-metal accumulators (LASAT, 2000, VASSILEV et al., 2004).

This generates a common debate about the choice of hyperaccumulator versus normal accumulator species. While the hyperaccumulator has the capacity for high metal accumulation, it is limited to slow growth and little biomass, and the normal accumulator has a low metal concentration which is compensated for by the greater volume of biomass. The degree of metal removal depends on the biomass harvested and the metal concentration accumulated (LASAT, 2000).

However, phytoremediation is not simply a matter of planting and cultivating a few metal-hyperaccumulating species in a given contaminated area. It refers to a highly complex and strategic procedure, which requires a project with professionals who have experience in the field, who will be able to select the appropriate species for specific metals, taking into account the peculiarities of the contaminated region (ALKORTA et al., 2004 *apud* GRATÁO et al., 2005).

Defining the species for phytoremediation is a very important factor in the efficiency of the metal removal process. Although extraction potential is important, there are other criteria that influence this choice. A native species is preferable to an exotic one, which can be invasive and threaten the harmony of the ecosystem. To avoid the spread of weeds, it is recommended to use cultivated plants, but the seeds and fruits can be an attractant for animals, putting them at risk

(LASAT, 2000).

As the vegetation grows, depending on the type of phytoremediation chosen and the type of contaminant, the biomass should be removed from the site and disposed of appropriately. Dead material such as fallen leaves, diseased or dead plants must be analyzed in order to control the accumulation of the contaminant in this material. If it is not contaminated with toxic elements, it can be used as fertilizer, otherwise it needs to be treated and disposed of properly, avoiding potential risks such as introduction into the food chain. Depending on the element extracted, the biomass can be used to replenish minerals in poor soils (EPA, 2000).

In the case of heavy metals, if it is present in the aerial part of the plant, it can be harvested using traditional farming methods. It is important to harvest before the leaves fall or the plant dies and decomposes, so that the contaminant does not return to the soil or contaminate other areas through dispersion. After harvesting, the biomass can be processed to extract and separate most of the metals (LAMEGO; VIDAL, 2007).

Thermal, physical, chemical or microbial processes can be used to reduce the volume and weight of biomass. The burning method can generate energy, which is an economically viable destination, and the residual metal can be recovered from the ashes and treated as ore (CUNNINGHAM; OW, 2010, GRATÀO et al. , 2005 *apud* LAMEGO; VIDAL, 2007).

2.51 NFLUENCE OF METALS ON PLANTS

In order to understand the effects of lead on plant organisms, it is necessary to review their nutrient process. This will be influenced by the availability of the metals in the soil, the way in which they are sorbed by the species and their action on the plants, differentiating essential, beneficial and/or toxic elements and identifying the symptoms of plant poisoning.

In the soil, metals are associated with other elements: (1) in the soil solution, with free metal ions and soluble complexes; (2) adsorbed to the inorganic parts that make up the soil, through ion exchange; (3) bound to the soil's organic matter; (4) precipitated as oxides, hydroxides and carbonates; (5) incorporated into the structure of silicate minerals. Bioavailability thus depends on the solubility of the metal, with only elements in states one and two being available for plant absorption (TESSIER et al., 1979 *apud* LASAT, 2000, RAIJ, 2011).

To define whether an element is essential or not for a given plant species, three criteria are used: analyzing whether the absence of the element is fatal to the plant before it completes its cycle; criterion 1: if there is a deficiency directly related to the lack of the element, criterion 2: it cannot be replaced by any other and criterion 3: when the element is part of a reaction that is essential for plant metabolism (MENGEL; KIRKBY, 1982, MARSCHNER, 1995).

All nutrients are taken up by plant organisms through root absorption, where an element enters the plant. It occurs through contact between the element and the root, and can take place through interception, when in the process of growth the plant root encounters the element; mass flow, when the element is solubilized in water and in the process of absorption it is absorbed together; and diffusion, when the element moves through the water in a stationary phase and is absorbed together (RAIJ, 2011).

Translocation, the movement of metal, takes place across membranes and is controlled by proteins with a transport function. In addition, sensitive mechanisms maintain the concentration of metal ions within physiological requirements. In general, the absorption mechanism is selective, with some being preferentially assimilated over others. This selection depends on the structure and properties of the transporter membrane, causing the transporters to recognize, bind and transport specific ions (LASAT. 2000).

Because of the metal charges, the metal cannot move freely across the cell membrane. For this to happen, ionic transport must be mediated by conductors that bind exclusively to the metals to be transported, connecting the extracellular part with the intracellular part and being receptive exclusively to the ion to be transported. Both mineral and organic components are transported in the same way. Heavy metals can be taken up as chelates in the transport process (LASAT. 2000).

Essential nutrients act in a specific way on plant development. They are classified as: (1) structural, when the element is part of the molecule of organic compounds; (2) enzyme constituent, being a form of structure that are usually metals or transition elements, which are part of the prosthetic group of enzymes and are necessary for their functioning; and (3) enzyme activator, when the element, even though it does not participate in the prosthetic group, is fundamental for its activity (MALAVOLTA, 2006).

Essential elements, when in high concentrations, can do just as much harm as toxic elements. Therefore, in response, common accumulator plants have developed absorption control mechanisms, including stimulation of transporter activity, inhibition at high concentrations and expulsion of the element back into the external solution (LASAT, 2000).

2.52 OXIDATION OF PLANTS BY LEAD

The absorption of lead is probably a passive process, i.e. non-metabolic, which takes place quickly, although there is little long-distance transport. The endodermis acts as a partial barrier, accumulating the metal. In the case of accumulator plants, absorption is progressively lower as the concentration of the medium rises. In the case of excluded species, the amount in the aerial part remains constant in the presence of a broad concentration range up to a maximum absorption limit and above that it is not regulated (MALAVOLTA, 2006).

Lead in its ionic form, Pb^{2+} , is not very mobile in the soil, but the organic forms of Pb-tetraethyl, triethyl and diethyl are extremely mobile and quickly reach plant roots (FAQUIN, 2005).

Toxic effects of lead have included changes such as chlorosis and blackening of the root system, inhibition of photosynthesis, alteration of the nutritional system for minerals, water balance and the hormonal system, as well as a reduction in plant development (an effect observed in studies using nutrient solutions) (SHARMA; DUBEY, 2005).

The absorbed Pb accumulates in the cell walls, which probably reduces its toxic effect and transport to other parts of the plant. Studies show that lead toxicity is related to iron metabolism and the formation of heme groups (FAQUIN, 2005).

Species with resistance to Pb toxicity love detoxification strategies such as sequestering the metal in the vacuole, phytochelatin synthesis, and the ability to restrict the metal to the cell wall (SHARMA; DUBEY, 2005).

Inorganic lead tends to form insoluble and complex salts, reducing its availability to plants via the roots, so for it to cause visible toxic effects on photosynthesis, growth and other parameters, it must occur at very high concentrations (WHO, 1995).

According to Who (1995), Bazzaz's 1992 experiment with *Helianthus annuus* (sunflower) plants subjected to a lead salt solution for five days showed a reduction in photosynthesis and respiration. The same author in another experiment tested the behavior of corn and soybeans in the presence of lead chloride, which also showed a reduction in photosynthesis and respiration.

The author also explains that Barker's experiment in 1972 exposed cauliflower stems, carrot roots, potato tubers and lettuce to lead acetate solution, resulting in a decrease in the fresh mass of lettuce and carrots, while cauliflower and potatoes showed a decrease in growth.

Who (1995) also mentions that the experiments carried out by Bell and Patterson in 1926 with *hyacinth* bulbs exposed to lead acetate solutions showed inhibition of root growth at the highest concentration.

The US *Environmental Protection Agency* (EPA) has been carrying out research into *phytoremediation* since the 1980s and presents a table of remediating plants taken from the *Introduction to Phytoremediation*. It lists plants that have been proven to remediate heavy metals, including *Brassica juncea,* popularly known as brown mustard, *Thlaspi caerulescens* and *Alyssum wulfenianum,* which belong to the Brassicaceae family; *Populus deltoides*, Populus *charkowiieensis* and *Populus tricocarpa,* hybrid trees belonging to the Salicaceae family; and *Helianthus annuus*, popularly known as sunflower.

Some research carried out in Brazil has been taken as a reference, because it verifies the lead remediation potential of some species. The experiments were carried out in the laboratory, in controlled environments, generally with solutions, as shown in Table 1.

Table 1 - Species used in Pb phytoremediation

Species	Popular name	Reference
Ricinus Communis L.	Castor beans	LIMA, 2000
Melissa officinalis	Lemon balm	CHAVES et al., 2010
Pteris sp.	Fern	AMARAL, 2013
Andropongon sp.	Grass	AMARAL, 2013
Helianthus annuus L.	Sunflower	LIMA, 2010, PEREIRA, 2005
Canavalia ensiformis	Pork beans	PEREIRA, 2005, ROMEIRO, 2007
Zea Mays	Corn	PEREIRA, 2005
Mentha crispa L.	Mint	SÁ et al, 2015
Macadamia integrifolia	Macadamia	VILAS BOAS et al, 2015
Jatropha curcas L.	Jatropha	GOUVEIA; MACRUZ; ARAÚJO, 2015
Mucuna aterrima	Black mucuna	SANTOS et al., 2013

2.53 SPECIES ADOPTED IN THE RESEARCH

Four plant species were chosen for this study to check their remediation potential in lead-contaminated soils. Some factors were decisive in the choice of species, such as morphology, the species not being edible, good adaptation to the region, rapid growth and low purchase and maintenance costs.

Braga (2012) explains that the choice of different plant morphologies helps in the composition of the landscape project proposed by urban planners and agricultural engineers for the recovery of contaminated areas. The characteristics presented by each species, such as a variety of colors, textures, shapes and volumes, help to make the site more visually interesting, as shown in Figure 01.

Figure 1: Fazenda Tacaruna - RJ Source: BURLE MAX (2016)

In order to create spaces of compositional interest, it is necessary to work in harmony and balance, which does not mean symmetry. Through rhythms, units, volumes, dominances and visual weights, it is possible to provide dynamism in the space and thus arouse greater interest and zeal in the population (CHING, 2006).

The species are not edible or have fruit that can be eaten by animals or humans, minimizing the possibility of contamination of the food chain if the contaminant is absorbed and translocated in the plant (EPA, 2000).

We selected from the book Plantas Ornamentais no Brasil: arbustivas, herbáceas e trepadeiras (Ornamental Plants in Brazil: Shrubs, Herbs and Climbers), by the authors Lorenzi and Souza (2001) species that currently make up Brazilian gardens and those that, while still in the wild, have the potential to do so, as well as being commonly found in urban landscaping in the region. In addition to being herbaceous species, they have poorly consistent tissues, which influences their rapid growth in order for them to respond quickly to contamination. Also considering the low cost of purchasing the seedlings T. pallida (R$1.70), O. jaburani (R$2.35), S. trifasciata (R$5.00) and C. gracilis (R$1.70) (from Big Garden Nursery) and low maintenance, this makes the experiment and its application on a large scale viable.

The species chosen and their main characteristics are described below in the alphabetical order of their families:

According to Figure 2, *Tradescantia pallida* belongs to the Commelinaceae family and is popularly known as purple ragweed and purple coraga. It is a herbaceous succulent native to Mexico and should be grown in full sun so that the color of the leaves is evident. It is widely used as fodder and in clumps, as a perennial plant with enriched soil and good fertility and organic material, kept moist and does not tolerate low winter temperatures (LORENZI; SOUZA, 2001).

Figure 2: *T. Pallida* - Central jardin of Av. Sao Paulo in Maringá-PR
Source: AUTHOR (2015)

According to Figure 3, the species *Ophiopongo jaburani* belongs to the Liliaceae family and is commonly known as snake's beard. A herbaceous plant with ornamental foliage, native to China and Japan, it can be grown in semi-shade or full sun in borders or as fodder. It likes soil rich in organic extract and permeable, and should be kept moist. It has leaves with golden-yellow stripes for decorative effect, and its inflorescence is in short spikes, with white flowers, which are not very ornamental (LORENZI; SOUZA, 2001).

Figure 3: *O. jaburani* - Unicesumar construction site in Maringá-PR Source: AUTHOR (2015)

According to Figure 4, the species *Sansevieria trifasciata* also belongs to the Liliaceae family, and its popular name is Sword-of-Saint-Jorge. This herbaceous plant of African origin can be kept in full sun or half shade in borders and massifs. It has good resistance to arid soils and tropical heat, as well as to cold, and because of this good adaptation it is found in various regions of Brazil. It has a wide variety of leaves with creamy-yellow margins, short, with transverse, grayish spots. Its inflorescence has no ornamental importance, but is long and spike-shaped with small white flowers (LORENZI; SOUZA, 2001).

Figure 4: *S. trifasciata* - Central median of Paraná Avenue in Maringá-PR Source: AUTHOR (2015)

According to Figure 5, the *Cuphea gracilis* species belongs to the Lythraceae family and is popularly known as erica or falsa-érica, a herbaceous perennial native to Brazil. It has small, evergreen leaves and small lilac or white flowers that bloom all year round. Suitable for forming borders and fodder plants, it appreciates half shade or full sun, with soil enriched with organic matter and good drainage and does not tolerate the cold (LORENZI; SOUZA, 2001).

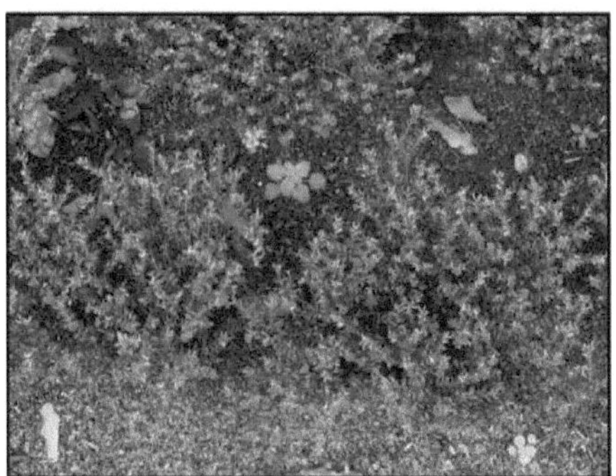

Figure 5: *C. gracilis* Central jardin of Av. Goiás in Cianorte-PR Source: AUTHOR (2015)

Chapter 3

3 METHODOLOGY

The methodologies studied as a basis for developing the experiment were those described by Chaves et al. (2010) and Pereira (2005).

Chaves et al. (2010) describe that the experiment took place in a greenhouse at the Agricultural Engineering Academic Unit of the Federal University of Campina Grande from July to October 2011, using plastic pots with a capacity of 10 L which were filled with 8.60 kg of substrate. The experiment was set up in a completely randomized design, with three replications, with five doses of Cu and Zn (0; 20; 40; 60 and 80 mg dm-3) making up a total of thirty experimental units, fifteen of which with Cu and another fifteen with Zn, the elements used as contaminants were copper sulphate and zinc sulphate.

The units received seeds of *Jatropha curcas L.* (pine nuts). After 30 days, one plant per unit was kept and the experiment lasted 90 days, monitored daily. After 90 days, the plants were harvested and the plant material was separated into roots, stems and leaves and dried in a forced-air oven at 70° C until constant weight, then weighed and ground in a Wiley-type mill. The material was then nitroperchloric digested to determine Cu and Zn in the extracts by atomic absorption spectrophotometry.

Pereira (2005) describes that the experiment took place in a hydroponic greenhouse at the Central Experimental Center of the Agronomic Institute in Campinas (SP) from March 26 to May 4, 2004. The treatments consisted of increasing concentrations of Pb (0, 100, 200 and 400 pinol L^{-1}) in a nutrient solution using lead acetate (Pb $(CH_3COO)_2.3H_2O$). The plants used were *Canavalia ensiformis L.* (hog bean). The experimental design adopted was entirely randomized blocks, in a 4 x 1 factorial scheme, with three replications, five plants per plot.

The seeds were left to germinate for six days and after seven days the plants were transferred to pots containing nutrient solution. After eleven days, the nutrient solution in the pots was replaced with a solution treated with increasing concentrations of lead. The plants were kept exposed to the metal for 28 days. At the end of the experiment, the aerial parts and roots were separated and dried in an oven at between 65 and 75 °C until they reached a constant mass. The material was then ground in a Willey mill for analysis of the Pb content. Dry digestion was carried out by incinerating the samples and dissolving the ashes in hydrochloric acid (HCl) and Pb was determined by plasma-induced flame spectrophotometry.

According to Abreu (1997), after the samples have been taken, it is necessary to interrupt the chemical and biological processes in order to preserve them from microbiological damage during the storage process. Enzymes are inactivated at temperatures above 60°C, which is why the material is dried in an oven until it reaches a constant weight.

To analyze the samples, they must be solubilized, i.e. the organic matter must be destroyed and the elements must be in solution. The methodology chosen for the process was the dry method, which involves incinerating the samples in a muffle furnace and dissolving the waste in an acidic medium (ABREU, 1997).

In this procedure, oxygen participates in the incineration process by oxidizing the organic matter, resulting in only inorganic waste. It is then dissolved in an acidic medium. This method makes it possible to work with a larger quantity of samples and better control contamination (NOGUEIRA, 2003).

Analytical methods require calibration of the reading equipment, and this process includes a calibration curve. Thus, several standards of known concentration of the material to be analyzed are introduced into the equipment, and the instrumental response is recorded. It is hoped that over a relevant concentration range a linear curve will be obtained, which is less prone to error. The equation is then adapted to the calibration curve using the least squares technique so that the sample concentrations can be calculated directly (SKOOG, 2002).

The research was carried out on the campus of Unicesumar, in the municipality of Maringá, in the state of Paraná, Brazil, with a latitude of 23° 26' South, a longitude of 51° 55' West and an altitude of 513 meters. It lasted 29 weeks, from March to October 2015.

The seedlings were obtained from the Big Garden nursery, located at Av. Petronio Portela, 1086 - Zona 38, Maringá - PR, were transplanted into plastic pots containing substrate (carbonized rice straw) and soil (eutrophic red latosol), collected in the 0 to 20 cm layer, in an uncultivated area of the Unicesumar university campus, and therefore with no history of application of agricultural chemicals. The mixture was thoroughly homogenized, moistened and placed in the pots, together with the seedlings, which remained in the greenhouse for a period of eight weeks for acclimatization (PEREIRA, 2005, CHAVES et al., 2010), as shown in Figure 6.

Figure 6: Vases in the Vegetagao House Source: AUTHOR (2015)

The experiment was carried out in a completely randomized design, with 4 replicates for each contaminated and uncontaminated case, totaling 8 pots per species, according to Table 2, divided into: I) samples without lead contamination, called witnesses and II) samples contaminated with lead, to which about 20.0 grams of lead sulphate II ($PbSO_4$), brand Vetec (content 99%, PM= 303.26 g/mol) was added to the soil.

Table 2 - Experimental design

Species	Trapoeraba purple	Snake beard	Sword of St. George	Erica	Quantity
Witnesses	4	4	4	4	16
Contamination (20g)	4	4	4	4	16
Total					32

Contamination with lead sulphate took place by inserting the solid into openings arranged laterally, as shown in Figures 7 and 8, close to the roots of the plants. The purpose of adding the solid to the top was to release the element Pb (lead) slowly, as a result of the irrigation. The aim is for small portions of the Pb ion^{2+} (toxic) to leach downwards, simulating the contamination of a solid waste dump. The amount deposited is well above what is established by the National Environmental Council (CONAMA) (2009), which recommends values of 200, 350 and 1,200 mg kg^{-1} for agricultural, residential and industrial areas, respectively.

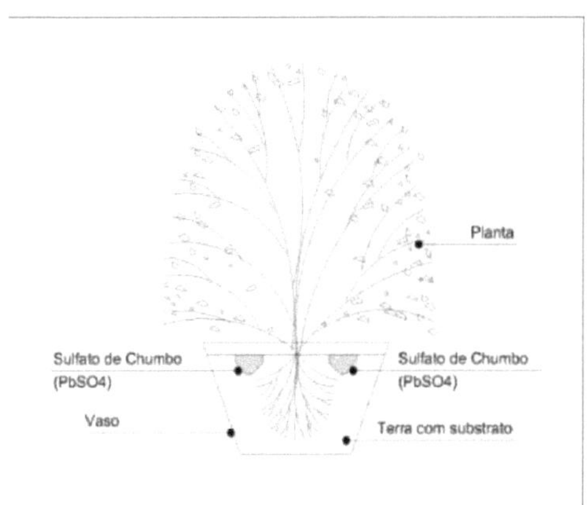

Figure 7: Illustration of contamination Source: AUTHOR (2015)

Figure 8: Lead sulphate contamination Source: AUTHOR (2015)

During the course of the experiment, weekly photographs were taken (at regular intervals of seven days) to monitor and compare the development of the contaminated plants with the control plants, observing signs of differences such as lower growth, necrosis, reduced photosynthesis, among others, according to experiments cited by Who (1995) and Faquin (2005).

After 20 weeks from the date of contamination, the aerial parts of the plants were harvested, their biomass reduced to smaller parts, packed in trays and dried in a De Leo oven at a temperature of between 65 and 70°C until they reached a constant weight, as shown in Figure 9.

After 27 weeks of contamination, the pots were dismantled and opened. In the process of cleaning the roots, it was observed that most of the material placed for contamination was still in the substrate, as shown in Figure 10, with little percolation. The substrate from the uncontaminated pots was discarded and those containing lead sulphate were separated for proper disposal.

Once dry, the samples were crushed into smaller pieces, placed in crucibles, weighed and taken to the SP Labor® muffle furnace with a maximum heating ramp of 1200° C, remaining at a temperature of 800°C for 4 hours and 1200°C for 8 hours, so that the organic part could be incinerated.

Figure 9: Greenhouse drying

Source: AUTHOR (2015)

Figure 10: Concentrated Lead Sulphate in the Substrate Source: AUTHOR (2015)

Calibration for lead quantification was carried out using standard solutions prepared with lead sulphate and distilled water in ten different dilutions, the most concentrated of which was the salt solubilization limit, 3.8mg/L, and the two smallest were disregarded as they were below the equipment's reading index, as shown in Figure 11.

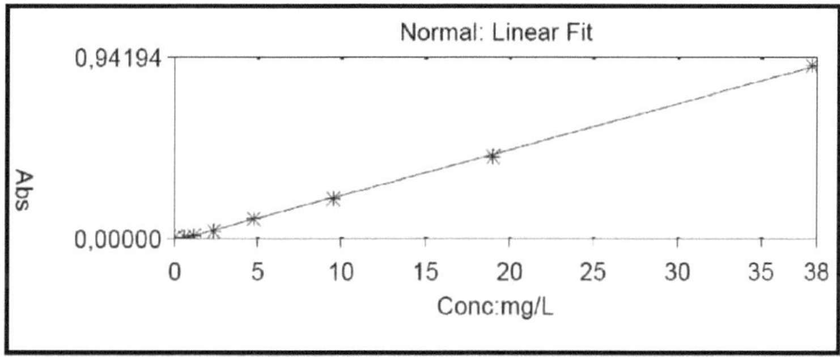

Figure 11: Lead Sulfate Standard Curve (Atomic Absorption Spectroscopy) Source: THERMO SCIENTIFIC (2015)

The analyses were carried out by flame and plasma atomic absorption spectroscopy at the Texsa laboratory, located at Av. Florivaldo R. Tampelini, 1470, Parque Industrial I, Umuarama/Paraná, by the chemist in charge, Djalma Palin Junior, on November 20, 2015.

Chapter 4
4 RESULTS AND DISCUSSION

The observations were made by comparing control and contaminated samples and the following aspects were considered:

1- Vertical and horizontal growth;
2- Foliage;
3- Flowering (when it fits);
4- New foliage appears;

The *T. Pallid* (purple ragweed) plants showed that the plants that received the contaminated soil developed normally compared to the control plants, as shown in Figures 12, 13 and 14. The chemical results did not indicate significant amounts of Pb in the aerial part of the contaminated soil plants.

Figure 12: *T. Pallida* - Witness Source: AUTHOR (2015)

Figure 13: *T. Pallida* - Contaminated with Lead (20.0g) Source: AUTHOR (2015)

The chemical results of the root analysis showed a quantity of 2.54mg of Pb /g of plant, proving the adsorptive capacity of this species, demonstrating an adsorption potential equivalent to approximately 0.176% of Pb per dry mass of its root.

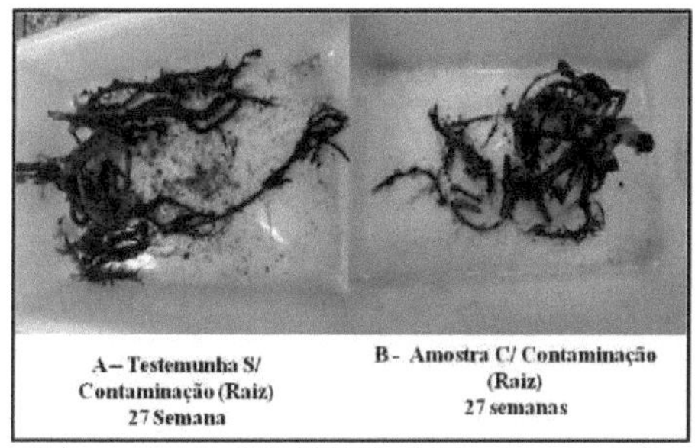

Figure 14: *T. Pallida* (root) Source: AUTHOR (2015)

The short incubation period does not allow for a better quantification of this potential, but the results demonstrate the possibility of using *T. pallid* as a Pb remediator. There were complications in handling this species when taking photographs, as it breaks very easily.

In the *O. jaburani* species (snake beard), it was possible to observe normal sprouting of new leaves in the contaminated plants in the same proportions as the control plants, as shown in Figures 15, 16 and 17.

Figure 15: *O. jaburani* - Witness Source: AUTHOR (2015)

Figure 16: *O. jaburani* - Contaminated with Lead (20.0g)
Source: AUTHOR (2015)

The chemical results of the roots showed a quantity of 0.91mg of Pb /g of plant, proving the adsorptive capacity of this species, showing an adsorption potential equivalent to 0.04% of Pb per dry mass of root.

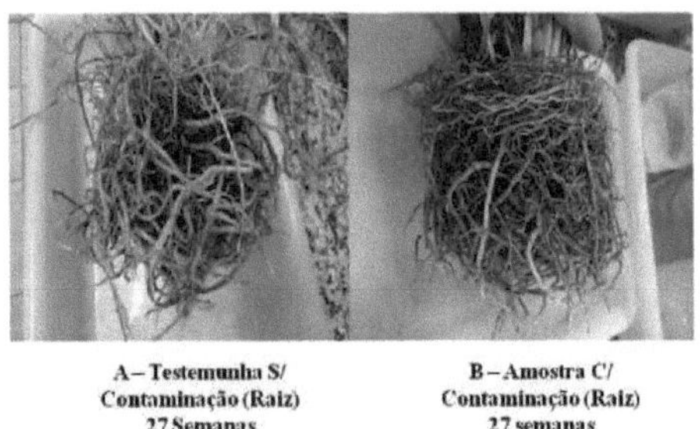

Figure 17: *O. jaburani* (root) Source: AUTHOR (2015)

The short incubation period does not allow for a better quantification of this potential, but the results demonstrate the possibility of using *O. jaburani* as a Pb remediator.

Individuals of the species *S. trifasciata* (St. George's Wreath) showed results that demonstrate greater sprouting and growth compared to the control plants, as shown in Figures 18 and 19.

As for the chemical results obtained from the leaf part, the absorptive capacity of this species was observed, as it showed a quantity of 0.75mg of Pb /gram of leaves, which is equivalent to 0.043% dry mass.

Figure 18: *S. trifasciata* - Witness
Source: AUTHOR (2015)

Figure 19: *S. trifasciata* - Contaminated with Lead (20.0g) Source: AUTHOR (2015)

The roots also showed greater development in terms of volume, as shown in Figure 20. The chemical results showed the absorption capacity of the roots, with 0.7879mg of Pb mass/mass retained, which is equivalent to 0.053% of the dry mass.

Figure 20: *S. trifasciata* (root) Source: AUTHOR (2015)

The short incubation period does not allow for a better quantification of this potential, but the results demonstrate the possibility of using *S. trifasciata* as a Pb remediator.

C. gracilis (erica or false erica) showed constant flowering on the plants that received Pb contamination, and accelerated sprouting in the aerial part, as shown in Figures 21 and 22.

Figure 21: *C. gracilis* - Witness Source: AUTHOR (2015)

The roots of the control plants visually compared with those that had been contaminated showed similar characteristics, as shown in Figure 23. As for the chemical results, a quantity of 1.3 mg Pb /gram of plant was observed, proving the adsorption capacity of this species, demonstrating an adsorption potential equivalent to 0.2% of Pb per dry mass of its root.

Figure 22: *C. gracilis* - Contaminated with Lead (20.0g) Source: AUTHOR (2015)

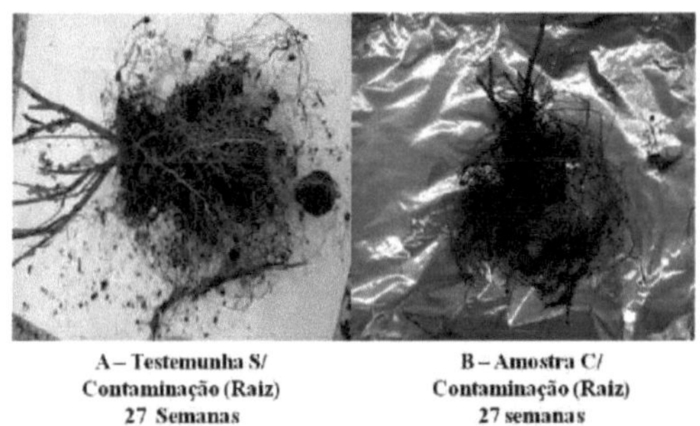

Figure 23: *C. gracilis* (root) Source: AUTHOR (2015)

The short incubation period does not allow for a better quantification of this potential, but the results demonstrate the possibility of using *C. gracilis* as a Pb remediator.

The species studied showed visual characteristics that contradict the literature reviewed, with no symptoms of intoxication such as reduced growth, leaf necrosis or inhibition of photosynthesis, as presented by Sharma and Dubey (2005).

The abundance of flowering shown by the species *C. gracilis*, popularly known as erica or false erica, may be indicative of an alteration in the hormonal system caused by intoxication, as presented by Sharma and Dubey (2005).

CONCLUSIONS

The species studied *Tradescantia pallida* (purple ragweed), *Ophiopongo jaburani* (snake beard), *Sansevieria trifasciata* (St. George's sword) and *Cuphea gracilis* (erica or false erica) showed potential for adsorbing Pb on their roots.

When visually assessing the growth of the species, comparing the uncontaminated and contaminated individuals, it was not possible to see any signs of poisoning, as they grew normally.

Through laboratory chemical analysis, it was possible to determine that all the species studied have a remediation potential in retaining Pb in their roots, depending on the way the contaminant was made available in the soil, with *T. pallid* showing an amount of 2.54 mg of Pb mass/mass retained, equivalent to 0.176% in the root, *O. jaburani showing an amount of approximately 0.91 mg of Pb mass/mass retained, equivalent to 0.04% in the root, S. trifasciata showing an amount of approximately 0.7879 mg of Pb mass/mass retained, equivalent to 0.*176% in the root. *jaburani had* approximately 0.91 mg of Pb mass/mass retained, equivalent to 0.04% in the root, *S. trifasciata* had 0.7879 mg of Pb mass/mass retained, equivalent to 0.053% in the root, and *C. gracilis had* 1.3 mg of Pb mass/mass retained, equivalent to 0.2% in the roots, as shown in Table 3.

Table 3 - Remediation potential of the species studied

Species	Pb m/m root	% at the root	Pb m/m Aerial
T. Pallid	2.54mg	0,176%	-
O. jaburani	0.91mg	0,04%	-
S. trifasciata	0.7879mg	0,053%	0.75mg
C. gracilis	1.3mg	0,200%	-

Only *S. trifasciata* showed translocation of Pb to the aerial part and was efficient at accumulating 0.043% of Pb, showing Pb phytoextraction potential. With the exception of S. *trifasciata*, plants develop defense mechanisms to prevent the contaminant from being translocated to the aerial part.

Among the species observed, *C. gracilis had the* highest Pb retention efficiency, equivalent to 0.2%, and *O. jaburani had the* lowest retention, 0.04%.

All the species studied that received Pb contamination showed a similar appearance to the control plants and none of the individuals died, indicating resistance to the element.

It should be noted that although the amount of Pb adsorbed and absorbed by the plants is low, this is due to the short exposure interval of the plants to ionizable Pb.

Suggestions for future studies include:

1- Determine the real absorption potential of *S. trifasciata* (espada-de-sao-jorge).

2- To study the metabolic process of *S. trifasciata* (St. George's wort).

3- Determination of plant absorption and adsorption in the presence of Pb^{2+} in solution (irrigation).

4- Check whether plants are able to adsorb/absorb more if exposed for longer to ionizable Pb.

REFERENCES

ABRELPE - ASSOCIAÇÃO BRASILEIRA DE EMPRESAS DE LIMPEZA PÚBLICA E RESÍDUOS ESPECIAIS. **Panorama of Solid Waste in Brazil 2013**. Available at: <http://www.abrelpe.org.br/Panorama/panorama2013.pdf>. Accessed on: May 4, 2015.

ABREU, M. F. **Simultaneous Extraction and Determination by Plasma Emission of Nutrients and Toxic Elements in Samples of Agronomic Interest**. 1997. 132f. (Doctoral Thesis) University of Campinas, Campinas, 1997.

AMARAL, D. C. **Ultrastructural Studies and Bioaccumulation Capacity of Zn, Cd and Pb by Plants in a Zinc Mining Area**. 2013. 99f. (Master's dissertation) in Soil Science, Federal University of Lavras, Lavras, 2013.

ATSDR - AGENCY FOR TOXIC SUBSTANCES AND DISEASE REGISTRY. ***Toxicological Profile ForLead.*** 2007. Available at: <http://www.atsdr.cdc.gov/toxprofiles/tp.asp?id=96>. Accessed on: May 4, 2015.

BAIRD, C.; CANN, M. **Environmental Chemistry**. 4ª ed. Porto Alegre: Bookman, 2011.

BENTO, J. C. C. V.; MENDONCA, N. A. A.; REIS, L. M. M. **Importancia da Remediacao e Recuperacao de Áreas Degradadas pela Indústria do Petróleo a Luz da Sustentabilidade**. VII CONNEPI-Congresso Norte Nordeste de Pesquisa e Inovagao, Palmas, Oct. 21, 2012.

BRAGA, M. A, (Coord.). **Municipal Landscape Resources Course**, 2012. Available at http://www.prefeitura.sp.gov.br/cidade/secretarias/upload/chamadas/recursos_paisagisticos_1431454341.pdf. Accessed on: May 5, 2015.

BRAZIL. Law No. 12.305, of August 2, 2010. Establishes the **National Solid Waste Policy**; Amends Law No. 9.605, of February 12, 1998; and makes other provisions. Presidencia da República - Casa Civil - Subchefia para Assuntos Jurídicos, Brasília, DF, Aug. 2, 2010. Available at: < http://www.planalto.gov.br/ccivil_03/_ato2007-2010/2010/lei/l12305.htm>. Accessed on: May 12, 2015.

BRUNA, P. J. V. **Arquitetura, Industrializacao e Desenvolvimento.** 2nd ed. Sao Paulo: Perspectiva, 2002.

BURLE MARX LANDSCAPING OFFICE. 2016. **Tacaruna Farm - RJ**. Available at: <http://burlemarx.com.br/inicio/>. Accessed on: January 28, 2016.

CATROGA, A. M. D. **Contribution to the Study of the Potential of Kenaf (*Hibiscus cannabinus* L.) in the Phytoremediation of Soils Contaminated with Heavy Metals**. 2009. 117p. (Master's dissertation) in Bioenergy - Hydrosphere Ecology Discipline Group, Faculty of Science and Technology, Universidade Nova de Lisboa, Lisbon, 2009.

CHAVES, L. H. G.; MESQUITA, E. F.; ARAUJO, D. L.; FRANCA, C. P. Accumulation and Distribution of Copper and Zinc in Papaya Cultivar BRS Paraguacu and Growth of PL. **Revista de Engenharia Ambiental**, Espirito Santo do Pinhal, v°. 7, n°. 3, p. 263-277, jul/set, 2010.

CHING. F. D. K. ; BINGGELi, C. **Illustrated Interior Architecture**. Translated by Alexndre Ferreira da Silva Salvaterra. 2ª ed. Porto Alegre: Bookman, 2006.

ENVIRONMENTAL COMPANY OF THE STATE OF SÄO PAULO - CETESB. Contaminated Areas Department. **CETESB Contaminated Areas Management Manual**. 2nd ed. Sao Paulo, 2001.

NATIONAL ENVIRONMENTAL COUNCIL (CONAMA). **Resolution No. 420, of December 28, 2009**. Provides for criteria and guideline values for soil quality in terms of the presence of chemical substances and establishes guidelines for the environmental management of areas contaminated by these substances as a result of anthropic activities. CONAMA, Brasília, DF, December 30, 2009. Available at: < http://www.mma.gov.br/port/conama/legiabre.cfm?codlegi=620>. Accessed on: May 11, 2015.

CONSTANTI, M.; MENEZES, E. Report Proves High Concentration of Heavy Metals in Dam Sludge. **R7 NOTÍCIAS**, Belo Horizonte, Nov. 2015. Minas Gerais section. Available at: <http://noticias.r7.com/minas-gerais/laudo-comprova- alta-concentracao-de-metais-pesados-em-lama-de-barragens-13112015>. Accessed on: Feb. 12, 2016.

COSTA, C. What is already known about the impact of the Marinana mud? **BBC Brasil**, Sao Paulo, Dec. 2015. News section. Available at :< http://www.bbc.com/portuguese/noticias/2015/12/151201_dados_mariana_cc>. Accessed on: 12 Feb. 2016.

CUNNINGHAM, S. D.; OW, W. D. *Promises and Prospects of Phytoremediation*. **Plant Physiolgy**, v° 110, n° 03, p. 715-719, jul./set, 2010. Available at: <http://www.ncbi.nlm.nih.gov/pmc/articles/PMC157769/>. Accessed on: Feb. 24, 2015.

DAMIANI, D. ; et al. Encephalopathies: etiology, pathophysiology and clinical management of some of the main forms of the disease. **Revista Brasileira Clínica Médica**, v° 11, n° 1, p. 67-74, 2013.

NATIONAL DEPARTMENT OF MINERAL PRODUCTION (DNPM). **Statistical Yearbook of the Metallurgical Sector. 2014**. Available at: <http://www.mme.gov.br/documents/10584/1865251/Anuario_Setor_Metalurgico_2014 _base_2013.pdf/136ee218-8fb3-430d-97dc-2e9171c6a0e8>. Accessed on: August 3, 2015.

FANTASTIC. 2015. **Inmetro Rejects Two Paint Brands with Lead Presence**. Available at: <http://g1.globo.com/fantastico/noticia/2015/10/inmetro- reprova-t duas-marcas-de-tintas-com-presenca-de-chumbo.html>. Accessed on: 02 Nov. 2015.

FAQUIN, V. **Nutricao Mineral de Plantas**. 2005, 175f. (Monografía) Post-Graduation in Soils and Environment - Foundation for Support to Teaching, Research and Extension - FAEPE, Federal University of Lavras - UFLA, Lavras, 2005.

FERNANDES, F. R. C. ; BERTOLINO, L. C. ; EGLER, S. G. Projeto Santo Amaro - BA: agglutinando ideias, construindo solugöes - diagnósticos. 2ª ed. Rio de Janeiro: CETEM/MCTI, 2012.

GLOEDEN, G. E. ; SOARES, M. S. **CETESB**, São Paulo, Dec. 2013. Directorate of Environmental Control and Licensing. Explanatory Text - List of Contaminated and Rehabilitated Areas in the State of São Paulo. Available at: <http://cetesb.sp.gov.br/areas-contaminadas/wp-content/uploads/sites/45/2013/11/texto- explanativo.pdf>. Accessed on: May 4, 2015.

GOUVEIA, A. F.; MACRUZ, P. D.; ARAÚJO, J. H. B. Phytoremediation of Lead-Contaminated Soils Using *Jatropha Curcas L.* ***Blucher Chemical Engineering Proceedings***, v° 1, n° 2, p. 8213-8219, feb, 2015.

GRATÄO, P. L. et al. *Phytoremediation: green technology for the clean up of toxic metals in the environment*. ***Brazilian Journal of Plant Physiology***, v°, 17, n° 1, p. 5364, jan./mar, 2005.

HOLZBACH, J. C., et al. Lead: an introduction to extraction and phytoremediation. ***Journal of Biotechnology and Biodiversity***, v° 3, n° 4, p. 178-183, nov., 2012.

BRAZILIAN INSTITUTE OF GEOGRAPHY AND STATISTICS (IBGE). **Population of Brazil**. 2015. Available at:< http://www.ibge.gov.br/apps/populacao/projecao/>. Accessed on: May 9, 2016.

NATIONAL NEWSPAPER. **Population Drinks Water Contaminated by Lead in City.** 2016. Available at:< http://g1.globo.com/jornal-nacional/noticia/2016/01/populacao-bebebe-agua-contaminada-com-chumbo-em-cidade-americana.html>. Accessed on: Feb. 12, 2016.

KHAN, A. et al. *Role of Plants, Mycorrhizae and Phytochelators in Heavy Metal Contaminated Land remediation*. **Chemosphere**, v ° 41, n° 1, p. 197-207, jan./mar, 2000.

LAMEGO, F. P.; VIDAL, R. A. Phytoremediation: plants as agents of depollution? **Revista de Ecotoxicologia e Meio Ambiente**, v° 17, n° s/n, p. 9-18, jan./dez. 2007.

LASAT, M. M. ***The Use of Plants for the Removal of Toxic Metals from Contaminated Soils***. 2000. Available at :< http://www.plantstress.com/articles/toxicity_m/phytoremed.pdf>. Accessed on: May 9, 2016.

LIMA, A. M. **Evaluation of the Phytoremediation Potential of Castor Bean (*Ricinus communis* L.) and Sunflower (*Helianthus annuus* L.) for the Removal of Lead and Toluene in Synthetic Effluents**. 2010. 110 f. (Doctoral Thesis) in Research and Development of Regional Technologies - Federal University of Rio Grande do Norte, Natal, 2010.

LOPES JR et al. **Lead and Arsenic in the Sediments of the Ribeira de Iguape River - SP/PR**. 2006. Available at:< http://www.cprm.gov.br/publique/media/geo_med14.pdf>. Accessed on: May 9, 2016.

LORENZI, H.; SOUZA, H. M. **Plantas Ornamentais no Brasil**: arbustivas, herbáceas e trepadeiras. 3ª ed. Nova Odessa: Instituto Plantarum, 2001.

MALAVOLTA, E. **Manual de Nutricão Mineral de Plantas**. 3rd ed. São Paulo: Agronómica Ceres, 2006.

MALIK, N.; BISWAS, A. K. *Role of Higher Plants in Remediation of Metal Contaminated Sites*. ***Scientific Reviews and Chemical Communications***, v° 2, n° 14, p. 146, 2012.

MARIANO, D. C.; OKUMURA, R. S. Agronomic Aspects, Human Use and Mechanisms of Phytoremediation: a review. **Rama - Revista em Agronomía e Meio Ambiente**, v° 5, n° 2, p. 85-101, Oct, 2012.

MARSCHNER, H. ***Mineral Nutrition of Higher Plants***. 2nd ed. San Diego: *Academic Press Limited*, 1995.

MENGEL, K.; KIRKBY, E. A. ***Principles of Plant Nutrition***. 3rd ed. Worblaufen-Bern: *International Potash Institute*, 1982.

MILHORANCE, F. The Health Risks of National Cosmetics. **Jornal o Globo**, Rio de Janeiro, June 2013. Society section. Available at: <http://oglobo.globo.com/sociedade/saude/os-riscos-saude-dos-

cosmeticos-nacionais- 8498885>. Accessed on: November 2, 2015.

MOREIRA, F. R.; MOREIRA, J. C. The Effects of Lead on the Human Organism and Its Significance for Health. ***Panamericana de Salud pública***, v° 15, n° 2, p. 119-129, jan, 2004.

NOGUEIRA, A. R. A. **Sample Preparation.** NATIONAL MEETING ON THE METHODS OF EMBRAPA LABORATORIES. Jaguariúna. New perspectives for Embrapa Laboratories. 2003.1CD-ROM.

OLIVEIRA, R. F.; GOMES, E. S.; AFONSO, J. C. O Lixo Eletrónico: uma abordagem para o ensino Fundamental e médio. **Química Nova na Escola**, v° 32, n° 4, p. 240-248, nov., 2010.

PANTAROTO, H. L.; FIGUEIREDO, P. J. M. **Lead**: exploration, use and public health. XIV Symposium on Production Engineering - SIMPEP, Bauru, November 9-11, 2007.

PAOLIELLO, M. M. B.; CHASIN, A. A. M.; (2001) - Ecotoxicology of Lead and its Compounds. **Série Cadernos de Referencia Ambiental**, v° 3, n° s/n, p. 144, 2001.

PEREIRA, B. F. F. **Potencial Fitorremediador das Culturas de Feijão de Porco, Girassol e Milho Cultivadas em Latossolo Vermelho Contaminado com Chumbo**, 2005. 68f. (Master's Dissertation) in Tropical and Subtropical Agriculture - Management of Agro-environmental Resources, Agronomic Institute, Campinas, 2005.

RAIJ, B. V. **Soil Fertility and Nutrient Management**. 1ª ed. São Paulo: Agronómica Ceres, 2011.

ROMEIRO, S. et al. Lead Absorption and Phytoremediation Potential of *Canavalia Ensiformes L.* **Bragantia**, v° 66, n° 2, p. 327-334, 2007.

SÁ, R. A. et al. ***Phytoaccumulation and Effect of Lead on Yield and Chemical Composition of Mentha Crispa Essential Oil.*** 2015. Available at: < http://www.tandfonline.com/doi/abs/10.1080/19443994.2013.874716>. Accessed on: May 9, 2016.

SANTOS, C. H. et al. Performance of Mucuna Preta (*Mucuna Aterrima* Piper & Tracy) in the Phytoremediation of Lead Contaminated Soil. **Revista Agro Ambiente Online**, v° 6, n° 3, p. 215-221, 2013. Available at: ?< http://revista.ufrr.br/index.php/agroambiente/article/view/705>. Accessed on: May 9, 2016.

SHARMA, P.; DUBEY, R. S. *Lead Toxicity in Plants*. ***Brazilina Journal of Plant Physiology***, v° 17, n° 1, p. 35-52, jan./mar, 2005.

SKOOG, D. A. **Principles of Instrumental Analysis**. 5th ed. Porto Alegre: Bookman, 2002.

SPIRO, T. G.; STIGLIANI, W. M. Química **Ambiental.** 2aed. São Paulo: Pearson Prentice Hall, 2009.

TRÖSTER, E. J. Plumbica Intoxication. **Revista Pediátrica (USP)**, v° 15, n° s/n, p. 4549, 1993.

United States Environmental Protection Agency (EPA). *Introduction to Phytoremediation*. 2000. Available at:< https://www.epa.gov/nscep>. Accessed on: May 9, 2016.

VASSILEV, A. et al. *The use of Plants for Remediation of Metal-contaminated Soils*. ***The Scientific World Journal***, v° 16, n° 4, p. 9-34, jan., 2004.

VILAS BOAS, N. et al. **Evaluation of Kinetic and Thermodynamic Parameters in Adsorption of Lead (Pb^{2+}) and Chromium (Cr^{3+}) by Chemically Modified Macadamia (*MacadamiaIintegrifolia*)**. 2015. Available at:< http://www.tandfonline.com/doi/abs/10.1080/19443994.2015.1085906?journalCode=td wt20>. Accessed on: May 9, 2016.

WORLD HEALTH ORGANIZATION (WHO). 1977. ***Environmental Health Criteria 3***. Available at: <http://www.inchem.org/documents/ehc/ehc/ehc003.htm>. Accessed on: October 7, 2015.

Environmental Health Criteria 165. 1995. Available at: <http://www.inchem.org/documents/ehc/ehc/ehc165.htm>. Accessed on: August 13, 2015.

Printed by Books on Demand GmbH, Norderstedt / Germany